Keeping Australian Native Stingless Bees

Bees that are stingless, disease free and require zero maintenance - How cool is that?

by Greg Coonan

NB

Northern Bee Books

Published in the United Kingdom by
Northern Bee Books,
Scout Bottom Farm,
Mytholmroyd,
West Yorkshire HX7 5JS
Tel: 01422 882751
Fax: 01422 886157
www.northernbeebooks.co.uk

ISBN 978-1-912271-77-1

Cover image **Image 1 by Geoffrey Dutton**

Design and artwork, DM Design and Print

Keeping Australian Native Stingless Bees

Bees that are stingless, disease free and require zero maintenance - How cool is that?

by Greg Coonan

Foreword

Native stingless bees are one of nature's most wonderful gifts. Our passion is native beekeeping, and we enjoy making it easy for others to gain the knowledge and confidence to look after and be enriched by these fabulous little creatures. This is the sole purpose for preparing this guide.

Based on my own years of experience and those of my friends I have prepared this simple guide for first time hive owners, and for those who want to progress into native beekeeping.

Our methods are based around those so rigorously researched, tested and wonderfully presented by Tim Heard in his book 'The Australian Native Bee Book'. This easy to read book contains a wealth of knowledge, and should be studied by anyone wishing to progress from owning a hive to native beekeeping, undertaking such practices as honey harvesting and hive propagation. The more you read and understand about the biology of native bees and in particular native stingless bees, the better prepared you will be to become a native beekeeper. As with any profession or trade, once you become proficient, you make small adjustments to the practices and procedures to what works consistently well for you, while maintaining the core principles - it can be the same for native beekeeping.

This guide sets out what works consistently well for us, and for which advice is constantly given to hundreds of now successful hive owners. It is divided into two parts. The first part sets out what you need to know to become a hive owner of *Tetragonula carbonaria*, *Tetragonula hockingsi* and *Austroplebeia australis*. The second part deals with more advanced native beekeeping practices for *Tetragonula carbonaria* and *Tetragonula hockingsi*.

Acknowledgements

I would like to acknowledge and express my appreciation for the assistance provided in preparing this document by Rick Ada, Geoffrey Dutton, Jenifer Pedler, Ray Stewart, Bernetta Billing, Ed Endicott and Steve Armstrong.

A big thankyou to those photographers who have graciously allowed me to use their images: Geoffrey Dutton, Daniel Ellis, Ed Endicott, Ray Stewart, Layla Coonan, Emma Sorensen, Steve Armstrong, Sondra Grainger and Alistair Grubb.

Table of Contents

Image 2 Tetragonula carbonaria (TC) native stingless bee *by Daniel Ellis.*

PART 1 - Your First Hive

This first part sets out what you need to know to become a hive owner of *Tetragonula carbonaria (TC), Tetragonula hockingsi (TH) and Austroplebeia australis (AA)*. They are stingless, disease free and once fully established, are close to being impregnable to pests. There is zero maintenance required for literally years. **See Image 3 and 4** (see below).

Image 3 and 4 from left to right A native stingless bee enjoying a sunflower by Ray Stewart.

Native Bee Hive Ownership and Beekeeping

- **Who should own a native beehive?**
- **The difference between owning a native bee hive and native beekeeping**
- **Knowing your bees**
- **Log hives**
- **Box design**

Who should own a native beehive?

▶ **Animal lovers** - As anyone who owns a native bee hive will tell you, they are a joy to watch. Native bees are terrific, self-sufficient pets with none of the usual holiday pet hassles, and to cap it off, unlike keeping honey bees there are no registration requirements, no hassles with neighbours, no stings to worry about and no ongoing costs.

▶ **Honey lovers** - they produce delicious honey, between half and one kilogram per year. The honey has a similar taste to honey bee honey but with a citrus twang a bit like someone has placed lemon rind in your honey. Native stingless bee honey also possesses strong antimicrobial properties equivalent to high potency Manuka Honey. Recent University of Queensland research is also indicating sugar found in native stingless bee honey has a lower GI which means it takes longer for the sugar to be absorbed into the blood stream, so there is not a spike in glucose that you get from other sugars.

▶ **Keen gardeners** - Native bees are fun to watch, great pollinators of fruit trees, vegetables and flowers. There is no better gift for the keen gardener, particularly when you have run out of creative gift ideas for that special friend or family member. As native bee hives live so long, they can become a point of intergenerational connection, that can be passed on from generation to generation making them ideal wedding, retirement and birthday gifts (Image 5).

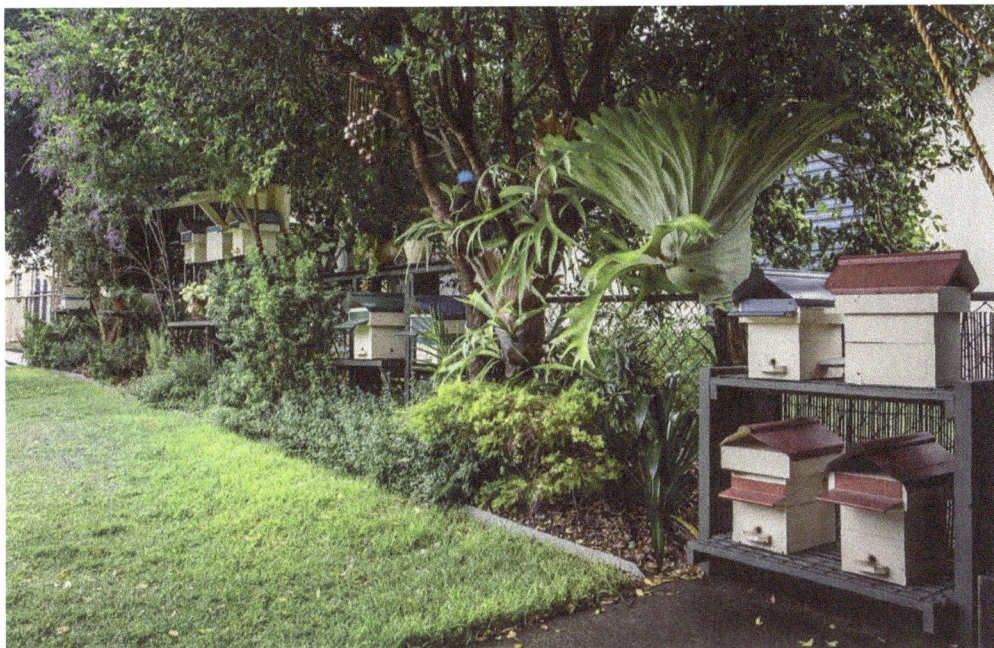

Image 5 *Native stingless beehives complimenting a garden* **by Layla Coonan.**

▶ **Retirees** - For a retiree or a person planning to retire, native beekeeping is a great hobby opportunity. As a hobby you can invest as little or as much time as you want, without being a drain on your finances as it is not too difficult for this hobby to pay for itself. Stingless bees are easy to manage, lightweight and do not take up much space.

▶ **Environmentally conscious** - For the person who simply wants to stay true to nature and have a positive impact on the environment, they can help pollinate the native plants and trees in their neighbourhood simply by owning a native beehive.

▶ **People with children** - Many families with young children want to introduce their children to owning a pet but may not have the confidence, or time, or be able to afford the costs normally associated with owning most pets, or do not want to be tied down with a pet at holiday time. A native bee hive is an ideal solution. Apart from the cost of initial purchase there are no ongoing costs, no registration fees and the hive is self-sufficient. Importantly, the children will find the frenetic activities of bees working from the hive educational and fun to watch, while being perfectly safe (Image 6).

Image 6 *Even babies and toddlers can be drawn to watching bees and are completely safe.*

The difference between owning a native bee hive and native beekeeping

▶ Owning a native bee hive is a pleasure and for many people an end in itself. From the people we deal with we would estimate that two thirds of hive owners are more than happy just owning a self-sufficient native bee hive and have no desire to harvest honey or to propagate further hives. Many people simply do not have the time or desire to gain the expertise to undertake honey harvesting, and hive propagation. As native stingless bees are disease free and, once fully established, are near impregnable to pests, a person who simply wants a native beehive can do so with zero maintenance required.

▶ However, for some people, the desire to develop their interest into a hobby sees them progress to native beekeeping, involving honey harvesting and/or hive propagation and/or pollination services. Many a native beekeeper also turn their hand to building their own boxes, roofs and stands and often provide hives as gifts to family and friends. A few of these native beekeepers progress to selling beehives and providing associated services, and they are usually completely addicted by this stage.

Knowing your bees

- The first thing you need to know about native stingless bees is that although they are social and live in a colony with a queen, they are quite different to European honey bees (*Apis mellifera*). There is much more to them than the inability to sting. Much of what you may understand about honey bees doesn't apply to native stingless bees. The biggest mistake we see people often make is they assume that native stingless bees can be purchased separately from a hive box just like honey bees can - particularly their queens. People need to buy already established native hives to get started - an empty box is of no use until you already have an established hive to propagate from.

- The now famous Flow Hives are for honey collection from European honey bees and cannot be used with native stingless bees.

- Native stingless bees are disease free and do not need to be split or have their honey harvested. Native stingless bees are good at self-regulating their food stores. (Note; while Queensland is disease free a bacterial disease known as Shanks Disease is present in some parts of New South Wales. It can be identified by the normally cream coloured brood cells having a blue tinge. Refer to Tim Heard's The Australian Native Bee Book chapter 14 for more detail.).

- There are three species of native stingless bees covered in this guide: -
  - *Tetragonula carbonaria* (TCs) – ranging from south east Queensland to the south coast of New South Wales,
  - *Tetragonula hockingsi* (THs) – ranging along the entire coast of Queensland and they are much smaller then European Honey Bees as shown in Image 7. The queens in each species have a much larger cream and brown banded abdomen and are about 1cm long.

- All three are black with grey heads and are stingless. TCs and AAs are only about 4 mm long and THs are slightly larger at about 4.5 mm long. The queens in each species have a much larger cream and brown banded abdomen and are about 1cm long (Image 7).

Image 7 Native stingless bees are much smaller than European Honey Bees (Honey bee on left side of flower).

▶ They are social bees and have a queen bee plus thousands of sterile female worker bees and some male drone bees in each hive. A mature TC and TH hive will hold about 8,000 to 10,000 bees with the queen laying around 300 to 400 eggs each day. AAs have about half as many bees, around 4,000 to 5,000. Once an egg is laid it takes about 50 days to reach maturity and a mature worker bee will live for about 100 days. Unlike European honey bee queens, the native bee queen is unable to fly again after she mates and enters a hive to begin laying. She has a much longer life expectancy then her worker bees of around 1 year. Because she cannot fly, she cannot swarm off with her bees to a new home like honey bee queens (Image 8 & 9).

Image 8 & 9 from left to right *Close up photo of a Tetragonula carbonaria (TC) queen. A Tetragonula hockingsi (TH) queen laying an egg in an open egg sac (Brood cell) surrounded by young nursery bees.*

▶ In the bush, stingless bees construct their nests in a cavity, usually the hollow in a tree left by termites. The bees use varying mixtures of wax that they secrete; and resins collected from plants, to create what is called propolis to build the structures of the hive. They will surround a central brood chamber with an involucrum sheath made up of multiple layers of propolis which is in turn surrounded by a complex array of pollen and honey pots within an interwoven propolis structure. Corridors are made throughout this maze of food chambers. Honey pots are recognised by their darker colour, while pollen pots are recognised by their paler colour. The walls of the hive are completely sealed with a dark resin mix called batumen.

▶ The TC brood mass is an amazing spiralling layered structure, cream in colour and perfectly proportioned. The queen lays eggs from bottom to top with a space known as the Advancing Front which separates the newly laid eggs from older hatching eggs. Each egg layer will have one or two larger queen egg cells (Image 10). The brood mass of AAs is not nearly as neatly structured as the other two species and can appear to be randomly placed in the hive (Image 11). Hockingsi brood masses are similar to TC brood masses with a layered egg structure but do not have a spiral structure or the spacing between Brood layers (Image 12).

Images 10, 11 and 12 from left to right Photos of Tetragonula carbonaria (TC),
Austroplebeia australis (AA) and Tetragonula hockingsi (TH) brood.

▶ Young bees commence duties in the nursery attending the needs of the queen, and possess a wax secreting enzyme that helps form the propolis building material. As the bees get older, they move into food and beverage by preparing the honey and pollen pots. The pollen is fermented with honey to produce digestible protein and of course, the honey provides the sugar energy. The next stage in a worker's life is guard duty at the hive entrance and then rubbish removal of mainly rolled up egg sacs. After that they enter the food gathering stage of their life cycle (Image 13 & 55). These older bees suffer a high attrition rate from predators and weather, hence the need for the queen to lay so many eggs. The final stage of their life cycle for those bees that make it this far, is being a scout for food and potential sites for a new colony.

Image 13 *Bee carrying pollen* **by Geoffrey Dutton.**

▶ The bees appear to be unionised as they don't like to work when it is too hot, too cold (to be fair to them they can't physically fly at less than about 17 degrees), too wet or too windy, and again to be fair to them, they are only about the size of a house fly. Don't be surprised if on difficult weather days that your hive may appear to have no bees, only to reappear when the weather is friendlier.

▶ They happily forage on native and exotic flowers to a range of about 500 metres.

▶ The bees carry pollen and resin for hive building on their hairy hind legs, and ingest nectar into their abdomen in a second stomach to carry it back to the hive to produce honey (Image 14 & 15).

Image 14 and 15 left to right Bees gathering resin from the exposed root of a Eucalyptus tree

▶ As well as a constant stream of bees bringing food and building material into the hive, there is usually a regular stream of bees removing rubbish from the hive (Image 16). This is most often pupal cases (rolled up egg sacs) and 'bee poo' in small baubles which they take well away from the hive. At the end of each day you will often observe bees dragging out bees that have died in the hive usually due to old age, and unceremoniously dropping them off the edge of the hive.

Image 16 spent brood waste being removed from hive **by Geoffrey Dutton**

- They have no ability to sting but have a bite a bit like a small black ant. The main defence when their hive is seriously threatened (i.e. physical interference - splitting hive, rescuing hive, transferring hive) is to try to annoy you by nipping your ears, nose and edges of your eyes. They will also crawl inside pants and shirts. We find it helpful to wear a fly net over our head when we need to work with them when they can fly, so that we can concentrate on the job at hand.
- Native bees and European honey bees happily coexist when foraging and if hives are located close together they will ignore each other.

Log Hives

▶ Log hives have been saved from being woodchipped or burnt by tree loppers, firewood cutters, fencing contractors and land clearers (approved clearing only). For us these are the most beautiful natural hives anyone could have. They make wonderful garden features with each hive being totally unique.

▶ For this reason, it is recommended that where ever possible, wild log hives are kept intact and not broken/cut open to transfer the hive into a box.

▶ We place hardwood floors and roofs on log hives to protect the hive from the weather and predators, and as a way of ensuring the integrity of the log itself for decades to come (Image 17 & 18).

▶ All a log owner needs is a suitable position and hard stand to place a log hive on - nothing more is required.

Image 17 and 18 left to right *Two log hives*

Box design

▶ Do bees care what their box looks like or what timber or other material is used to build the box? The answer is no. Bees need a chamber that they can secure and have sufficient volume (about 8 to 10 litres) for the hive to function fully.

▶ While the bees do not care what their box looks like, from a beekeeper's point of view, the box needs to be well made to provide good protection from pests, and to support consistent bee management practices with the minimum of fuss. Most beekeepers end up with standardised box designs but there are many different designs being adopted. The nearest thing to a national standard for box design is the OATH (Original Australian Trigona Hive) developed by Tim Heard. We follow and recommend the OATH standard design. OATH boxes have a base measuring 200mm x 280mm.

▶ A bee box comprises three sections;

  • The bottom section (90 - 100 mm high) has a drain hole in it, and an entrance hole. We add a doorstep at the entrance, not because the bees need one, but for the enjoyment of hive owners allowing them to more easily observe the antics and comings and goings of bees.

  • The middle section (90 - 100mm high) has a rebated top to accommodate a Perspex viewing panel/separator to contain the top of the brood mass to this section. There is a breather hole drilled into the rear of this section assisting the bees to circulate air through the hive.

  • The top section (about 50 mm high) comprising the honey super.

▶ The OATH hives are easy to make and there are also numerous suppliers in the market place. Using a standard sized hive means that a new hive owner is not tied to the hive supplier if they ever want a new box to split their hive, and they can buy new boxes for competitive prices.

▶ There are several types of timber that can be used to build the boxes. The density of the timber is an important consideration. Low density timber like hoop pine is a good insulator as it does not conduct heat as readily as say, very dense hardwood. When the weather is really cold, the hive does not want its heat to escape and when it really hot, they do not want heat conducted into the hive. If you want to use hoop pine 25 – 30mm thickness is recommended but it can be difficult to source. If you use the more readily available denser cypress pine you need to increase the thickness to about 35 – 40mm. If you want to use hardwood you need to increase the thickness to about 45mm to achieve the same level of insulation as the hoop pine. If thicker timbers are used it will reduce the internal volume to less than optimal. To offset this, a taller (100mm) and therefore larger, honey super can be made. To extend the box life boxes need at least two coats of quality exterior paint (Image 19).

Image 19 OATH dimension three tier hive boxes.

▸ To prolong the life of your box, and to help cool the hive in hot weather, we recommend the use of a roof which provides an airspace at the top of the hive. We use pitched timber roofs using hardwood ply or even better marine grade ply (Image 20).

Image 20 OATH box timber roofs

Hive Installation and Management

- **Tips for installing a boxed hive**
- **Tips for installing a log hive**
- **Problems to be alert to**
- **Hive relocation**
- **Co-location of hives**
- **Gentle Warm Day Swarming**
- **Brutal Invasion Swarming**
- **Keen Mating Swarming**

Tips for installing a boxed hive

▶ We have found owners of hives gain a great deal of pleasure from watching the activities of the hive and are usually thankful when they have placed them in a location close to where they relax in outdoor areas such as patios. A friend often comments that he feels like he gets a dollar back every time he sits down for a cup of tea to watch his bees!

▶ While ease of viewing can be important, it is more important to try to find a location with good shade in the summer months with little or no direct sun, while in winter enjoying good warming sun for most of the day. This is usually achieved by placing the hive under a shady tree on the northern or eastern side which the winter sun is able to get under to warm the hive. A north facing verandah or patio can work equally as well but the roof needs to be insulated to protect the hive from heat (Image 21 & 22).

Image 21 and 22 left to right Hives on a north facing verandah and front yard.

▶ If you don't have an ideal year-round location you can consider moving them short distances to take advantage of more sunlight or less sunlight depending on the season. Placing the hive on a small stool or table will make moving the hive an easy task. Moving needs to occur incrementally by about one third of a metre each day allowing the bees to reorientate themselves each time until a new location is reached.

▸ We avoid areas that are heat sinks such as concreted areas and uninsulated hot verandahs, and areas that are permanently shaded in winter such the southern side of a wall.

▸ If you are is still in doubt about where to place a hive, we suggest they think about where you would sit in your yard on a stinking hot summers day. If you like a particular spot, chances are the bees will also like that location (similar for very cold winter days).

▸ Desirably, the hive should face North or East so that the hive has its back facing cold south westerly winds and summer storms fronts - but this is not essential.

▸ During severe heatwaves hives can overheat and effectively melt killing the hive. You need to be sure that your hive is well shaded and has a good roof. A damp towel placed over the hive with one end in a bucket of cool water should be considered in the most severe of heatwaves. In hot weather some hives will form lines of bees on the front of the hive spiralling out of the entrance to draft air into the hive to cool it down (Image 23). At milder times of year, the bees also use this technique to draft air through the hive possibly to reduce the water content of their honey.

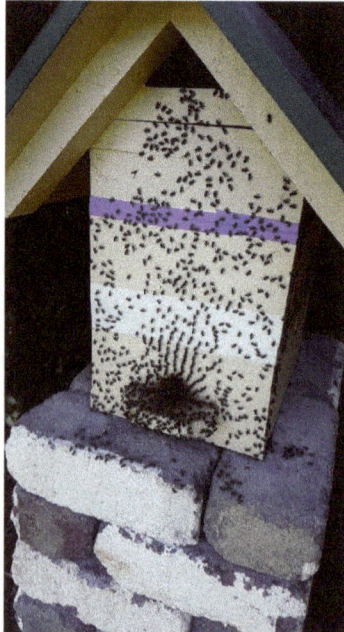

Image 23 Bees organised to cool a hive during heatwave conditions *by Ed Endicott.*

- Standing your hive can take many forms and can be adapted to the particular circumstances and opportunities in the yard the hive is going into. They can be placed in your garden, on patios, verandahs and balconies. The hive can be placed on a shelf, on a wall or fence, or on poles such as a post, a star picket or mail box stand. Another option is to place them on a small outdoor table or chair or a specially constructed platform made from materials like concrete blocks.
- For hives placed on a flat surface like a shelf you need to protect against moisture and potential rot in the floor of your hive so it is important to ensure that an air space exists between the hive and the surface it sits on. You can use thick wire mesh or insert protruding screws or studs on the bottom of the hive box. Separation also reduces the risk of white ant invasion of your box timber.
- The bees need ready access to sunlight - this is how the bees tell which way is up. Placing the bees at the rear of a wide covered verandah (closest to your home) will likely cause the hive to bunker down and not forage, which over time will lead to failure.
- The flight path of your bees should not be over a water body like a swimming pool. The sun reflecting off a pool surface can confuse the bees (as the bees use the sun to know which way is up) and result in them flying into the pool and drowning.
- Green houses are not a suitable location for hives as the bees tend to seek the sun when leaving a hive, resulting in them gathering under the roof of the greenhouse and unable to undertake food gathering.

If your hive is located in suburbia, it is paradise for them with something flowering all year round allowing many hives to thrive from a small yard. No supplementary feeding is required and in suburbia a water source is usually not required.

Tips for installing a log hive

- A log hive can make a wonderful addition to any garden.
- Log hives can provide a central feature that immediately attracts attention or can be very discreet and almost invisible (Image 24, 25 & 26).
- Just as every yard and garden is unique so are log hives. Log hives come in

all shapes and sizes, skinny and wide, short and tall and in varying timber colours depending on age and exposure to the elements (including charcoal from the effects of bush fires) but for the most part they are varying shades of grey.

▶ Even if you are fortunate enough to come by a log that does not seem to be the best fit, most owners find they grow on them because of the innate beauty of a prosperous hive.

▶ For a newly rescued log hive needing the top and/or bottom secured, we recommend use of hardwood decking timber over these openings. If you need to seal any gaps or cracks to protect the hive use a non-toxic filler like No More Gaps. **Liquid Nails should be avoided as it is toxic for the bees.**

▶ Some locations already give a firm base such as a paved or concreted area. In other locations we prefer to place concrete pavers or blocks on the ground to give a stable platform and to keep moisture and termites away from the log timber.

▶ Most logs, once provided with a level base, require no other support. However small, taller and unbalanced logs require support such as a discretely placed star picket or a tie to an adjoining fence or wall.

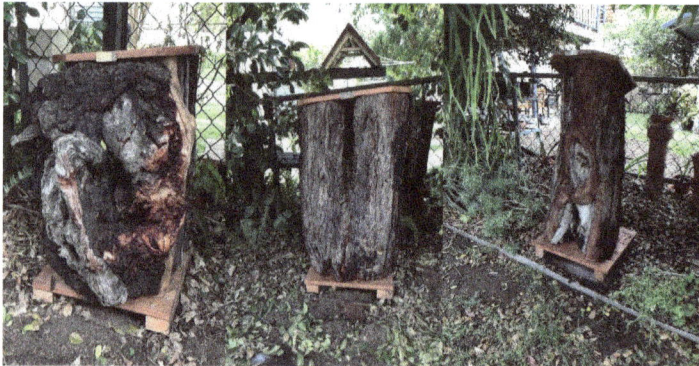

Images 24, 25 and 26 Left to right *Log hives in the garden*

▶ Logs can be safely transported lying down. Standing a log hive that was found lying on the ground does not cause a problem as the bees readily adapt.

▶ Suitable logs also make great mailbox stands and bases for garden features such as bench seats and tables, provided there is no seating directly in front of the hive entrance.

▶ You really can be creative and imaginative when installing a log hive in your garden (Image 27).

Image 27 Log hive with personality – by proud log hive owner Emma Sorensen

Problems to be aware of

▶ Insecticides should not be used close to native bee hives. Bees are small insects and are extremely susceptible to insecticides. People sometimes forget this when they are spraying their gardens and homes. Sometimes your neighbours will want to do the same thing without realising – it is worth discussing this problem with them. If you do need to use insecticide close to your hive where spray drift may occur, consider moving your hive to a friend's place for a couple of days, or at least cover the hive with an old sheet and spray in still conditions at night.

▶ Ant powders can also be a problem if spread on the ground in front of the hive where the workers drop to the ground with rubbish from the hive. If they land in this powder, they will pick up small traces of powder and bring it back into the hive. This can become toxic to the whole hive.

▶ Ants are commonly found around bee hives which is usually not a problem for mature hives as the bees defences are pretty much impregnable. The ants and bees happily coexist and can be ignored, but for rescued hives and newly split hives in their early, more vulnerable stages, it is best to keep the ants away from the hive by placing Vaseline on the hive support e.g. mail box stand.

▶ A range of predators eat bees or harvest them to feed their young, this is a natural process that is hard to avoid and for the most part has minimal impact on the hive. If spiders build webs in the flight path of the bees it is best to move the spiders on to a better location. Asian geckoes will sometimes take up residence near or on your hive - they are usually too quick to be caught. In warmer months, sand wasps (Bembix Wasps) (that resemble leaf cutter bees) are very clever at harvesting bees just before they enter, or as they are leaving the hive, by paralysing them and taking them back to a tunnel in the sand in which they lay their egg. Again, this is a natural process that the hive can handle. Another common predator are Assassin Bugs which catch bees with their proboscis, but again they are no real threat to the hive (Image 28 & 29). **Do not use insect sprays on ants, spiders, geckoes and wasps.** From time to time small insect eating birds such as silver eye may hunt some bees but they can be ignored as they do not eat many.

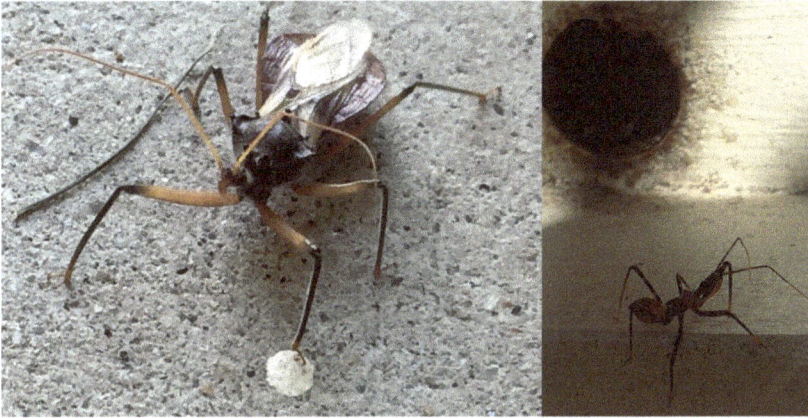

Image 28 and 29 left to right Assassin Bug

▶ Hives need protective barriers around them in exercise areas for more active physical family pets, and kids play areas, to stop impacts from footballs and cricket balls. Locations near vehicle turning and reversing areas need to be carefully thought about.

▶ African Tulip Trees are considered a low threat to native bees. Bees that enter the flowers of these trees don't seem to come out but the bees are not strongly attracted to this plant. After a number of years of keeping some hives near these trees it does not appear to have any measurable effect on overall hive health.

▶ Cadaghi trees are usually not a serious problem. Cadaghi is a native of North Queensland but in past decades it became a popular landscaping tree in South East Queensland and North East New South Wales. It flowers in November and gum nuts open in December/January. Native stingless bees seem to become addicted to gathering resin from the gum nuts for building. It was long thought that the Cadaghi resin had a lower melting point than other resins and that this could result in a hive slumping and failing during a heatwave, but scientific testing has shown this to be incorrect. Together with the resin, the bees also inadvertently bring back Cadaghi seeds that were attached to the resin. These commonly pile up at the hive entrance. Fortunately, the bees do not allow these seeds to interfere with the internal workings of their hive. However, there is a real risk for newly split hives and immature hives if they are placed near a large Cadaghi source. The bees can quickly fill the unused space in the hive box with Cadaghi seeds limiting the long-term development of the hive. In these environments it is best to split in early spring or early autumn. Another option is to ask a friend to bee-sit your hive during December/January. Hives with small entrance holes and those fitted with right angle bends, seem to be more prone to seed blockages.

Hive relocation

▶ Relocation sounds simple but if not done carefully, can take a heavy toll on your bee population. Your bees have a homing capability that allows them to navigate back to their hive after food gathering and rubbish dumping with great precision from 500 meters away. Even a short move of your hive for as little as 1 metre can result in your gatherers (around 20% of your hive numbers) being unable to find the new location and eventually perishing. Your hive may not die as the younger hive bees will eventually age and take their place as gatherers, but it does leave the hive more vulnerable.

▶ If the planned move is over a relatively short distance of a few metres, you can do this by moving the hive in small incremental steps of about 40cm each day until the new position is reached. The bees are able to reorientate themselves provided the daily incremental movement to the new location are small.

▶ However, for greater distances, the best method is to find a friend or relative, who lives more than one kilometre away, to bee-sit the hive for at least three weeks. If they are moved less than one kilometre, there is a risk that they will find old markers in their original 500 metre foraging range, that will lead them back to the original position of the hive. A move of greater than 1 kilometre allows the bees to be completely reorientated to the new location before bringing them back to your yard and placing them in a new position.

▶ Moving a hive from your yard is simple. When the bees have settled into the hive for the night they can be penned in ready for their move. To pen them in you can fold up a small piece of CHUX cloth and push it into the entry hole which blocks the bees while allowing the hive to breathe. A rolled piece of shade cloth can be used in the same way. We use a piece of cloth tape with a hole cut in the middle and small piece of fly screen placed over this hole. The tape is then placed over the entry hole with the flyscreened hole directly in front of the entry. Once the bees are penned in, they can be relocated that night or any time the next day. They can be let out on the first night at their new location. If you want to bring them back to your yard repeat the penning in procedure after they have been away for about three weeks.

Co-location of hives

▸ The suburbs have such an abundance of food that many hives can be housed in a backyard. A half dozen hives would thrive without having any impact on each other - some enthusiasts have many more.

▸ Hives from the same tribe e.g. two hives created by a split or a new educted daughter hive and a mother hive can often be kept together without any problems. But sometimes these hives will start to squabble, particularly when one of the hives is requeening. If they start to squabble, in time this will develop into fighting and if nothing is done, the weaker of the two hives will fail. The simple solution is to move one of the hives away.

▸ It is more difficult when combining hives from different tribes as they will be less tolerant of each other. Provided they are not too close and do not have flight paths that cross over, they seem happy to coexist. It is better to keep these hives 1.5 to 2 metres apart where you have bushes or trees between them, or 4 to 6 metres apart where there is nothing between them.

Gentle Warm Day Swarming

▸ Native stingless bee hives swarm for a number of reasons, but unlike honey bees, they do not swarm off to a new location as their queen cannot fly.

▸ On a sunny warm day in winter the bees will often show their pleasure with these friendly conditions to gently swarm near their hive. Wherever there is warm sunshine they almost float in the air only to return to their hive as soon as it cools down.

▸ At other times the hive will swarm around the hive more aggressively - usually after they have had a visit from another hive. As this could be a scout from a larger hive, the hive puts up a defensive swarm showing its strength. After a couple of days if no more visitors have arrived, the hive will settle down and get on with business as usual.

Brutal Invasion Swarming

▶ However, sometimes that unwanted visitor from another hive can lead to an invasion and massive fighting swarms. Fighting swarms are a different proposition to those mentioned above and you will be able to quickly spot the difference when fighting bees fall to the ground bound together in very large numbers. The local bee tribes in your neighbourhood could be all around without you realising as they are very discreet. These hives could be mature and looking for new nest sites to expand into. With so few suitable old hollow trees in suburbia the bees are adaptable and find many different homes such as water meter boxes and upside down flower pots, but sometimes it's your bee hive they become interested in as it is already stocked with supplies and 'ready to go'.

▶ We have found an attacking swarm can occur anytime other than the winter. It is a war of attrition and numbers. An invasion is usually preceded by 2 to 3 days of minor skirmishes usually on the door step of the hive to be invaded. As the bees enter into battle, they latch onto each other becoming locked together and falling to the ground until they die leaving a highly visible carpet of bees. Each morning the invaders arrive and a swirling mass of bees develops as bees lock together and fall to the ground. You will also notice in the flying and fighting frenzy that many individual bees simply fall to the ground exhausted, but after some rest on a blade of grass will re-join the fighting. This can go on for days or even more than a week until the hive with the most bees wins out.

▶ It can be distressing, but the insect world is quite brutal and the most difficult stage for a hive owner now occurs with the invading bees entering the hive and dragging out all the younger bees in the hive. The invading hive will then move in with a new queen and everything will settle down - often resulting in an even stronger hive. The invaders inevitably win the war.

▶ If you can intervene early enough, you can use a bait hive to seek to capture the invading swarm, please refer to 'Hive invasion capture' for details.

Keen Mating Swarming

▶ The most common swarm hive owners will witness is a drone swarm. When a hive needs to requeen, drones from up to 5 kilometres away will arrive at the hive and swarm around the hive during the day. As these drones are from different hives, they are not allowed to enter the hive with the guard bees killing any that try. This can result in a few dozen dead bees on the ground in front of the hive, but nothing like the carpet of dead bees left by fighting swarms. At night the drones will gather on nearby branches, twigs and leaves clustering together for mutual protection.

▶ At other times the drones will find a gathering point around an arbitrary structure, such as a post, that is up to 50 meters away from the requeening hive, indicating that the queen will come to them (Image 30).

▶ As the drones are unable to enter the hive of the new queen and do not appear to be able to return to their originating hive, they can swarm for days sometimes weeks around the hive of interest. There is no point trying to lure the drones into an empty hive box as they have no capacity to create a new colony.

Image 30 After a hard day of swarming these drones are settling in for the night on a small twig near a requeening hive.

Establishing a native bee friendly garden

▶ Your beehive will be looking for food sources and good shelter. A hive needs a location with good shade in the summer months with little or no direct sun, while in winter enjoying good warming sun for most of the day. This is usually achieved by placing the hive under a shady tree on the northern or eastern side which the winter sun is able to get under to warm the hive. Alternatively, deciduous trees like Crepe Myrtle and Frangipani can provide an ideal combination of summer shade and winter sun.

▶ As your hive has a foraging range of 500 metres and is a versatile feeder of both exotics and native plants it is important to get an appreciation of your neighbourhood's environment. If you are close to bushland that is a monoculture of say melaleuca or a streetscape dominated by say Golden Penda there is not much point planting the same plants in your garden. Unless perhaps you are trying to provide connectivity between two bushland areas.

▶ Plant a variety of plants to achieve a year-round food supply and cluster particular plants, this will attract more bees then scattered individual plants. Also look for different colours shapes and sizes bees are attracted to purple, blue, white and sometimes yellow coloured flowers. Bees seem to prefer flowers with a single layer of petals rather than multiple layers (Image 31, 32 & 33).

▶ Be wary of highly hybridised plants unless particular ones are recommended to you.

▶ We find locally owned nurseries and garden centres are usually the best place to obtain advice and to source your plants. Many will have native bee hives on display and staff members will often own a native beehive. This means they are in a position nearly every day to observe what plants the bees are foraging on and therefore to give the best advice.

▶ The plants listed in the tables 1 and 2 below attract foraging native stingless bees and most will attract native solitary bees. While the list contains plenty of reliable plant options it is by no means exhaustive and there will be many other options available in your local area.

Exotics (Table 1)

Flowers, herbs and ground covers	• Nasturtium *(Tropaeolum)* • Gazania (several varieties) • Catmint or Nepeta (many varieties) • Coneflower (Echinacea) • Bacopa *(Sutera cordata)* • African Daisy *(Osteospermum)*, Seaside Daisy *(Erigeron)* • Basil, Lavender, Rosemary Coriander, Borage
Small shrubs - 0.5 to 2 metres	• Lavender Lights or Texas Sage (Leucophylla) • Salvia (many varieties) • Sunflower (Helianthus) • Butterfly bush (Buddleja)
Medium shrubs – 3 to 5 metres	• Crepe Myrtle (Lagerstroemia) • Camellia (Many varieties)
Shade and feature trees – over 5 metres	• New Zealand Christmas Bush (Metrosideros)
Palms	• (Queen or Cocus Palms)

Natives (Table 2)

Flowers and ground covers	• Blue Eyes or Blue Sapphire (*Evolvulus*) • Cut Leaf Daisy (*Brachyscome*) and the Everlasting Daisy (*Bracteantha*) • Fan flower (*Scaevola aemula*)
Small shrubs – 0.5 to 2 metres	• Bottlebrush (Callistemon) • Grevillea (Grevillea) • Native lasiandra (Melastoma affine) • Tea tree (Leptospermum) • Grasstree (Xanthorrhea) • Pink Passion (*Eremophila maculate*) (Image 88)
Medium shrubs – 3 to 5 metres	• Tea tree (*Leptospermum*) • Lillypilly (*Syzgium*) • Purple pea bush (*Hovea acutifolia*)
Shade and feature trees – over 5 metres	• Golden Penda (*Xanthostemon chrysthus*) • Grevillea (*Grevillea sp.*) • Lemon scented myrtle (*Backhousia citriodora*) • Lillypilly (*Syzgium*) • Tulipwood (*Harpulia pendula*) • Gum trees (*Eucalyptus*) (*Corymbia*) (*Angophora*) Tea-trees (*Meleleuca*)

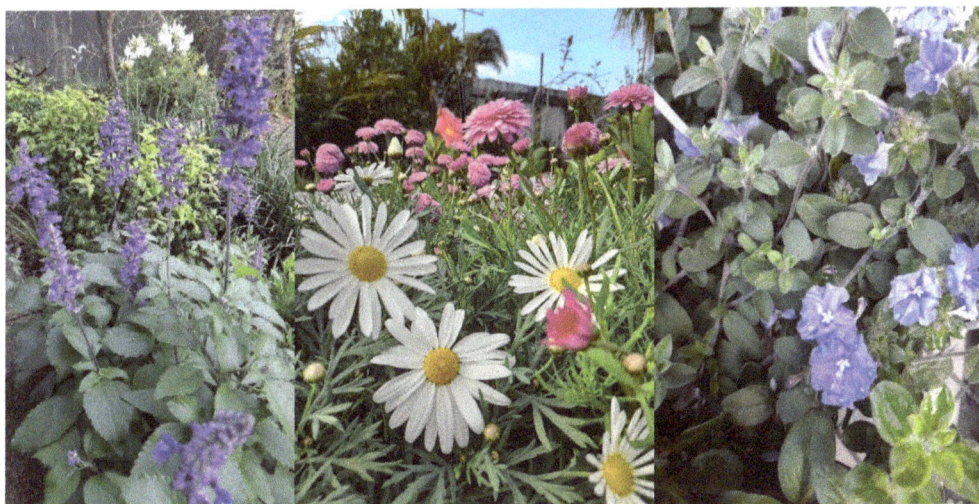

Images 31,32 and 33 left to right by Sondra Grainger

Ten hives we would never part with or propagate from and why they are special

▸ **Hive 1 Front Verandah TC Log Hive** – This beautiful small log hive sits on our front verandah and is admired by our visitors. It is very easy for small children to get a good look at the bees going about their business while looking down on their entry (Image 34 & 35).

Images 34 and 35 left to right

▸ **Hive 2 The Owl TC Log Hive** – This impressive old hive which greets everyone coming through our front gate with stunning owl-like eyes. We estimate the hive could be as much as 50 years old due to the large resin build up at the hive entrance (Image 36 & 37).

Images 36 and 37 left to right

▶ **Hive 3 Upside Down Pot/Fairy House TH Hive** – An upside-down garden pot occupied by a wild hive with a fairy house placed on top to amuse grandchildren. The fairy house is fully occupied by the bees giving them a two-storey home (Image 38, 39 & 40).

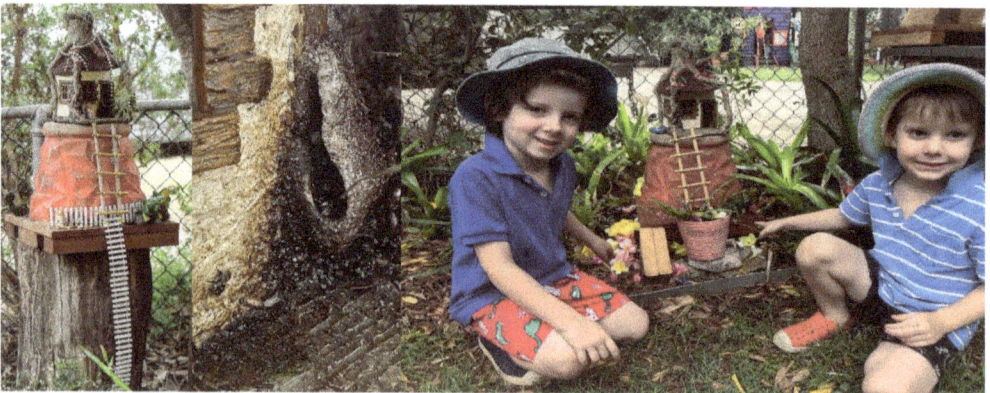

Images 38, 39 and 40 left to right

▶ **Hive 4 Water Meter Box TH Hive.** The hive was reburied in our garden to provide necessary insulation and fairy house placed on top with entry/exit hole (Image 41 & 42).

Images 41 and 42 left to right

▶ **Hive 5 Double AA Log Hive –** Log hives are always special to us so having a single log with two AA hives located in it is extra special (Image 43 & 44).

Images 43 and 44 left to right

▶ **Hive 6 The Beast TC Log Hive** – This is our biggest and heaviest log hive and the top of the log looks like the neck of a bullock, hence its nick name (Image 45 & 46).

Images 45 and 46 left to right

▶ **Hive 7 Double TC Log Hive** – Another wonderful log with two hives in it (Image 47, 48 & 49).

Images 47,48 and 49 left to right

▶ **Hive 8 Didgeridoo AA log Hive** – This hive resembles a didgeridoo in shape, size and internal dimensions. With the bees lining the pipe of the log with propolis, as they do, one wonders if thousands of years ago an indigenous Australian found something similar to this log and started a path to a new musical instrument (Image 50).

Image 50

▶ **Hive 9 AA Fishy Hive** – This hive colonised a ceramic garden ornament, a perfect home for many generations to come (Image 51 & 52).

Images 51 and 52 left to right

Hive 10 TH Log Hive Redcliffe Garden Centre – This is our only TH log hive and is on display at the Redcliffe Garden Centre (Image 53 & 54).

Image 53 and 54 left to right

PART 2 – Native Beekeeping

Part 2 covers native beekeeping processes using OATH boxes for Tetragonula Carbonaria and Tetragonula Hockingsi including propagation and honey harvesting but not commercial pollination services. We do not consider ourselves to be sufficiently experienced with Austroplebeia Australis to offer the same type of advice for this species.

Image 55 by Daniel Ellis.

Core Practices

- **Working with bees when they are not flying**
- **Working with reasonable speed**
- **Working with good box preparation**
- **Working with observation hoods**

Working with bees when they are not flying

▸ Outside of the winter months, we undertake beekeeping processes wherever possible when the bees are all in the hive and unwilling to fly. By doing it this way, we do not need a veil for rescues, splits and honey harvesting even for the somewhat more defensively aggressive THs. It is easier on the bees and the person undertaking the process. Fewer bees are lost in the process and the person undertaking the process is able to concentrate more fully on the task at hand producing better results. Anyone who has opened a TH hive on a warm sunny day will tell you just how aggressive and determined they are to get you to go away and just how hard it is to concentrate even with a veil on, leading to haste and mistakes.

▸ In early spring and late autumn, we look for morning temperatures around 13 to 17 degrees. On mornings above those temperatures, which usually applies to the whole of summer, we will undertake a process about 15 minutes before sunrise when there is enough light to see but not enough for the bees to fly with any precision.

▸ Of course, with rescues there are times when there is no choice but to transfer the bees when they are flying and you definitely need to wear a veil. It is also realised that commercial operators may not be able to only operate in such limited windows of time.

Working with reasonable speed

▸ To reduce the loss of bees from a beekeeping process, we find that moving quickly but not hastily, will reduce the number of bees lost. Leaving a hive open for any extended period, perhaps to satisfy someone's curiosity, results in many more bees leaving the hive to try to defend it and leaves the door open to pests to enter the hive.

▸ Good preparation and planning are key components of all processes. Unwanted delays can happen when all the materials and tools needed for the process are not readily to hand when you need them.

▸ It is helpful to have another person to assist you with a beekeeping process. An extra pair of hands and eyes will usually make the process quicker and easier and is a great way for an inexperienced person to gain the experience and confidence to undertake such processes themselves.

Working with good box preparation

▶ Before undertaking a beekeeping process good box preparation will improve your success rate.

▶ Our standard box preparation comprises:

- A perspex viewing panel rebated into the top of the middle box, which initially allows monitoring progress of the hive. When the hive is mature enough this viewing panel converts in function to become a honey separator. This is achieved by removing tape placed over a gap at one end of the perspex panel allowing the bees to access the honey super.

- A 5mm breather hole at the rear of the middle box for air circulation and a 5mm drain hole in the rear of the bottom box in case a drain is needed for leaking honey. Both holes are taped over with cloth tape that has a small hole cut into it with a slightly larger piece of fly screen placed under this, under the tape and against the hole. This prevents the entry of pests through these holes (Image 56).

- A flat piece of propolis placed over the inside of the entry hole with a small 2mm hole pushed through it. This gives the bees a defensive point until they are fully established. It also makes it easier to pen the bees in with tape on the outside of the entry hole which still has a clean surface (Image 57).

Images 56 and 57 left to right
Perspex viewing panel, rear breather hole protected with tape and fly screen and propolis protection inside the entry hole.

▶ In addition, for eduction purposes;

  • A hole drilled into one of the walls of the bottom box to insert a hose connecting this empty box to an intended mother hive;

  • Tape over the join between the bottom and middle sections of the eduction box to seal the box. The bees will also seal this joint from the inside with propolis as their first step in occupying the box. Once the join has been sealed and the propolis has hardened, the tape can be removed. This propolis usually takes about three months to harden (Image 58).

Image 58 *Taped box joins.*

▶ In addition, for splitting purposes;

  • Slump bars, in the form of masking tape, added to the top of the bottom box. Slump bars are added to avoid the contents of a full top box slumping into the new bottom box after splitting. Another option is to place permanent slump bars made of metal, perspex or timber in the bottom of the middle box. Because we do a lot of rescues, we have found permanent slump bars to be a hindrance to transferring hives into a box, hence we prefer to use tape (Image 59).

- Masking tape to join the honey super to the middle box to help with temperature control of the main chamber by the bees. It can be removed regularly to allow viewing of the progress of the hive and taped back on after each viewing.

Image 59 *Masking tape slump bars.*

▶ In addition, for hive rescue purposes;
- Wire mesh placed in the bottom of the box to keep the transferred hive out of leaking honey and to allow excess honey to more easily drain.
- A second layer of tape placed over the drain hole so that honey is not leaking out while the hive is being transported.
- Tape with fly screen placed over the entry hole to prevent the bees from escaping and to prevent pests from entering the hive.
- The two main boxes securely taped together so that no pests can gain entry. The perspex viewing panel is lifted to transfer rescued bees into the box (Image 60).

Image 60 *Rescue box ready for transfer with entry hole taped over to ensure the bees are penned in after transfer*

Working with observation hoods

▶ To permit easy monitoring of hives after splitting and rescue and during and after eduction processes, we use insulated timber hoods that slide over the middle box containing the perspex viewing panel.

▶ The honey super is put aside when the hoods are in use.

▶ These hoods are easily lifted off to allow ready viewing of the bees without disturbing them (Image 61 & 62). Any issues with the hive can be quickly identified and corrective action taken. Being able to observe the bees going about their business in the hive, including watching the queen and nursery bees from time to time, is both highly enjoyable and educational. After a couple of months the bees will cloud over the perspex as they seem to like their privacy.

Image 61 and 62 left to right Insulated viewing hoods

Propagation and Honey Harvesting

- **Do I have to propagate another hive from my hive and do I need to harvest its honey?**
- **When is your hive ready to propagate?**
- **Hive splitting**
- **Hive eduction – Natural**
- **Hive eduction – Assisted**
- **Hive eduction - Revival**
- **Hive invasion capture**
- **Honey Harvesting**

Do I have to propagate another hive from my hive and do I need to harvest its honey?

▶ Propagation refers to the generation of a new hive using splitting or eduction techniques. If you don't want to propagate from your hive and/or don't want to harvest honey from your hive then relax, you don't need to - its optional. Wild hives in logs survive for decades without any honey harvesting or splitting. We have also observed decades old hives in boxes that have never been split or had their honey harvested and they continue to thrive to this day. Native bees are not only disease free but are very good at self-regulating their food stores.

▶ An advantage of propagating at least one additional hive is that it acts like an insurance policy - in the event that a hive is lost, you still have a hive. As with honey bees, while there is a low risk of losing a mature hive there is always the chance of loss. Propagating a second hive provides insurance against such loss.

When is your hive ready to propagate?

▶ In Brisbane, TC and TH hives can be safely split on an annual basis following an initial split. But if you obtained your hive from a rescue, eduction or swarm capture your hive is likely to have a smaller volume than a newly split hive and may take about 18 months to 2 years to reach the same level of maturity. A more precise approach is to use the weight of the net contents of your box to decide. When your hive reaches a net weight of 4 to 4.5 kilograms it is usually ready to split. Our standard three tier boxes (25mm hoop pine) weigh 4.3 kilograms. The gross weight of a mature hive in a standard box is about 8.5 to 9 kilograms.

▶ When seeking to educt from a box hive or log hive the connection to an empty eduction box can happen at any time and the bees will naturally propagate when they are ready. But this will usually be when the hive is mature and starting to run out of space - including the honey super space which adds about another kilogram to overall hive weight when full. The available space within a log hive is unknown therefore it is difficult to predict when they are ready to educt but when they start, they can be prolific eductors as they are usually very strong.

Hive splitting

▶ We have consistently found hive splitting to be a safe and reliable method of producing two strong hives from a fully mature hive.

▶ Hive splitting involves dividing a mature hive into two parts. The mature hive is separated at the joint between the bottom and middle sections of the box. The full top section of the mature hive is placed on a new empty bottom box. A new empty top section is placed on the full bottom section of the mature hive. This process can be repeated roughly every twelve months. As with honey harvesting, some hive owners find this to be stressful because a number of bees usually die during the process. However, if you are prepared, quick, calm, methodical and split when the temperature is cool and bees are calm, the losses are minimal.

▸ With good preparation the process is quick, only taking about 2 to 3 minutes, and effective, producing two strong hives.

▸ The hive is split in two using a knife to cut the inner resin seal and then a splitting tool to lever the boxes open (Image 63). Before separating completely, it is important to make sure there is a fair amount of brood mass in each half (Image 64). If the entire brood is remaining in one half, this indicates that the rising front where the queen is laying, is either near the top or the bottom. For an inexperienced person it is better to rejoin the boxes and to repeat the splitting process in 3 or 4 weeks' time when the advancing front is closer to the middle of the box. With more experience it is not too difficult to cut between the egg layers of TCs and cut through the egg layers of a TH brood mass.

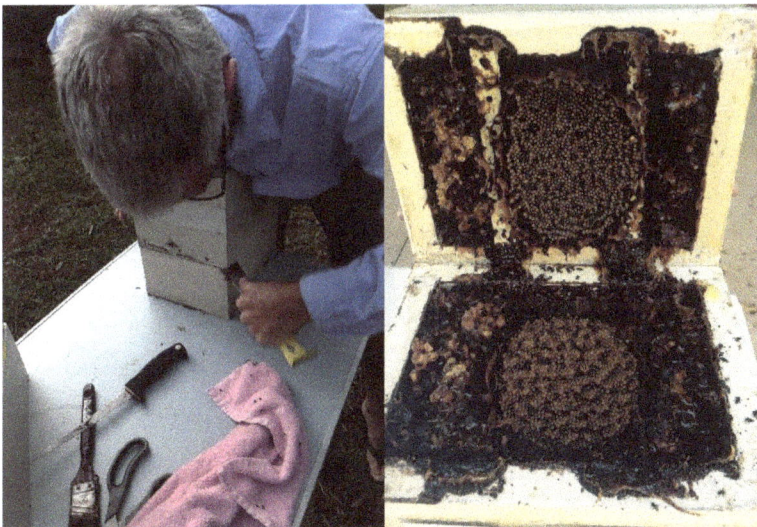

Images 63 and 64 left to right *Using a hive tool to prise open the join between the bottom two boxes. A newly split Tetragonula hockingsi hive ready to apply the new boxes to each half.*

▸ Make sure any honey that has leaked onto the outside of the box is carefully cleaned with a damp cloth to avoid attracting pests.

▶ Once the boxes are together, the joins of both hives should be taped over to keep pests out. The tape can be removed after about 3 months by which time the propolis seal made by the bees will have sufficiently hardened.

▶ The box with the new bottom is the most vulnerable to attack by pests and slumping. This box should be placed in the original position of the hive to ensure good bee numbers as the bees naturally return to this spot. This helps with defense against pests. We also place a propolis or Blu-Tack ring around the entrance hole, with a small 2 to 3 mm hole through it, to give the bees a smaller entry point to defend. If you do want to move a box away, the old bottom box with a new top is less fragile making it better for transport, and it inherits the protected entrance of the old hive. Both the split hives can be kept in the same yard or even together side by side.

▶ A problem we have encountered with hives located in much colder and hotter locations and with hives that have much larger internal dimensions then an OATH box, is that the bees can decide to locate their entire brood in one half of the box only. When this happens it effectively rules out the splitting option. Propagation is still an option using eduction.

▶ Splitting a hive that is not fully mature, even when the brood is fully developed, creates substantial risk of failure for one half, if not both halves, and should be avoided.

▶ Harvesting honey at the same time as splitting a hive creates a high risk of failure for the top half split. There will be a lot more bees lost and a lot of damage done for the remaining bees to repair. There is also usually a fair bit of leaking honey - much more so then a standard split, that will attract pests. We recommend honey harvesting and splitting processes be kept about 6 months apart to allow the hive to fully recover from each process.

▶ If you are splitting a hive for the first time because a friend or family member wants a hive, hold onto both splits for four months to be sure both are performing strongly before parting with one. This is so you are certain to have a hive yourself should one half fail which, however unlikely, is still possible.

Hive eduction - Natural

▶ Hive eduction, sometimes referred to as budding, involves connecting an empty box to a mature/strong hive, usually by a section of hose which forces the workers from the mother hive to gain access through the entry hole of the eduction box. The aim is for the mother hive to establish a new colony in the empty eduction box (Image 65, 66 & 67).

▶ Once connected, the mother hive will hopefully see the empty box as an opportunity to expand and develop a new colony. The mother hive will initially start to treat the empty box as a spare room and seal box joins with propolis to secure the chamber - including a defensive structure around the entry hole. The next stage will see a large number of bees occupy and take ownership of the eduction box - including building and provisioning pollen and honey pots.

▶ You can now wait for the bees to do their natural thing producing a new queen and starting a new colony. Sometimes you need to be patient as the queening process can take a few months to a couple of years. Sometimes the mother hive will simply use the eduction box for food storage and not allow a new colony to become established.

▶ Once the mother hive has taken ownership of the eduction box, you can encourage the mother hive to allow a new queen to establish by partially disconnecting the boxes. This gives the mother hive bees two points of entry. Partial separation can be achieved by pulling boxes apart until the connecting hose is just touching the wall of the eduction box allowing bees to enter through there or through the eduction box entry. Another way of achieving a double point of entry is to open a T piece inserted in the connecting pipe.

▶ It is important to be alert to the development of a brood mass which signals that a new queen is in occupation. As soon as we become aware of a new queen, we disconnect the hives and turn the mother hive in a different direction so that each hive has its own flight path. On several occasions when we had left the hives connected for much longer, hoping to build up the eduction box, we have lost or nearly lost the mother hive. The mother hive had become queenless and had become a food storage area only.

Our theories are that, either the mother hive queen decided the eduction box was a better location for her and moved over, or when the mother hive needed to requeen while still connected, it could not get a new queen past the queen now in the eduction box

▶ While the natural eduction process requires more time and effort than annual splitting, mainly in the form of monitoring progress, it can be an extremely rewarding the first time you see a new queen in your eduction hive.

▶ An eduction process can work equally well on a box or log hive but it is usually not possible to tell when a log hive is full to capacity and therefore more likely to colonise an eduction box.

▶ It is fun to watch the eduction process evolve by looking through the perspex.

Images 65, 66 and 67 left to right Box to box eduction - Log to box eduction - Box to log eduction.

Hive eduction - Assisted

▶ If you wish to accelerate the process, you can assist by taking some brood from the mother hive or another hive of the same species, and place it in the eduction box. The mother hive almost instantly takes ownership of the eduction box. In most instances this will result in the establishment of a new colony within a couple of months. If it hasn't worked the first time you can repeat the process until you get another hive established. We partially separate the boxes about one week after transferring the brood.

▶ Assisted eductions can occur a couple of times a year from a single mother hive. As with any process involving the opening of a hive, there will be a loss of bees associated with removing and transferring brood.

- Educted hives - natural and assisted - usually have a smaller volume than a newly split hive. Winter is risky time for small hives as the smaller mass and lower bee numbers mean there is less capacity to control temperature inside the hive.
- We find brood transfers between late August and November give the best opportunity for a new hive to establish itself, build some size and then get through its first winter. New eductions that get started close to winter are at a much higher risk of failure during their first winter.

Hive eduction – Revival

- We have found eduction techniques to be a valuable tool in reviving weak hives
- Sometimes with very small rescues and extremely difficult rescues, where not much can be salvaged or there is a lot of damage, there is a high risk of failure. However, if this recued material is placed into an eduction box connected to a strong mother hive it will nearly always result in the production of a new hive.
- We usually give these types of rescues a chance to make a go of it by themselves but as soon as it becomes evident there is further weakening the eduction process is put in place.
- We find that once a hive is showing signs of weakness there is no fight in the bees and they can be connected to a mother hive without any fighting. The mother hive within a few days, will pump a few thousand bee reinforcements into the eduction box.
- At other times hives simply fail to requeen for whatever reason - including instances where they have been left connected to an eduction box for too long. An eduction process is an effective way of reviving these hives.

Hive invasion - Capture

- An invasion creates a good opportunity to capture a new hive with a bait box. However, you need to become aware of an invasion within the first day or two. When it becomes evident your hive no longer has the numbers to continue the fight, it is too late to move the hive. Moving a severely wounded hive that has been under attack can result in the loss of that hive.

We leave a badly wounded hive to its fate knowing that at least we will still end up with a hive, probably a stronger one, and that it is a natural process, however brutal.

▶ Depending on the circumstances at the time, three options for a bait box can be considered.

- 1. If a mother hive with an eduction box connected is being invaded, you have a ready-made bait box. Simply disconnect and move the mother hive well away (more than 1 Kilometre) and leave the eduction box in position as the bait hive. This produces a new hive nearly every time.

- 2. A weak hive that may have been under consideration for eduction support, is also an ideal bait hive. Again, chances of a hive capture are high.

- 3. The most common option is to use an empty box that has spare propolis placed around the entry hole to try and convince the invaders it is still the hive they had targeted. In this case pen the hive in that is under attack and place the bait box under it. Keeping the hive that was under attack in that position keeps the invaders interested. After 24 to 48 hours move the penned in hive well away. This method has a reasonable chance of success and is well worth the effort.

▶ The invading bees will establish a functioning hive very quickly in a matter of 3 to 4 weeks. Once the invading hive is fully committed to building in the bait box you can bring your original hive back into your yard - but not to the same location as this will generate an unwanted fighting swarm. It is better to keep them about 10 metres apart.

Honey harvesting

▶ If your hive has a honey super with a separator in place, you have the option of harvesting honey when the super is full (Image 68). But there is no need to harvest this honey if you don't want to.

▶ Honey cannot be harvested from log hives and boxed hives that do not have a honey super and/or separator.

▶ Honey harvesting involves removal of the honey super (the top section of the hive) and pricking the honey pots with a tool that looks like a small bed of nails to allow the honey to be drained and sieved into a container (Image 69). Some bees will be caught and can drown in the honey.

- The honey is sieved to remove pollen and resin impurities and any bees that have been caught in the honey. We then do a fine sieve through muslin cloth.

- Pollen is a natural health food and a small amount in the honey is not a problem. Pollen in the honey will make the honey colour darker – the more pollen the darker.

- The honey harvest will be maximised with repeated pricking of the honey pots for about one minute and a long drainage period of about five minutes. Close to one litre of honey can be obtained. There is often a significant loss of bees during this type of extraction process.

- Two options are available to significantly reduce the loss of bees. By reducing the pricking process to about 15 seconds and draining the honey for about one minute the honey can be sieved while the bees have not yet drowned. The honey can then be gently hosed off the bees and they can be placed on the grass in the sun to dry, allowing them to return to the hive as soon as they can fly again. The compromise is a significantly reduced honey harvest of about 500ml. We think it is good compromise and the honey is clearer.

- Another method to reduce bee losses is to place the hive in a freezer for about two minutes before harvesting the honey. Just as happens on a cold winter's morning, the bees retreat to the inner brood chamber to help keep the brood warm. The honey super is left with just a few bees rather than hundreds.

- Once the honey super is placed back in position it is important that any honey leakage is wiped away with a damp cloth and the box join has tape placed around it to prevent any pests from accessing the hive via the honey leaks (where pest eggs can be laid). Check that the bottom drain hole is clear to allow any excess honey to drain if required.

Images 68 and 69 left to right A full honey super.
Piercing the honey pots to allow the honey to be drained away.

Hive Rescue and Managing Hive Pests

- **What is hive rescue?**
- **When can the hive stay? When must the hive be moved?**
- **Hive rescue by relocation**
- **When does a rescue hive need to be transferred into a box?**
- **Water meter box rescues**
- **Log hive rescues**
- **Hive box replacement**
- **Managing hive pests**

What is hive rescue?

▶ Hive rescue involves intervening to give a hive that will otherwise die or need to be exterminated, a chance to live. It is not about finding and removing bees from their natural habitat. As with other native creatures, native bee hives located in the natural environment of hollow trees should be left alone and not disturbed. For us transferring a hive into a box is never an automatic decision.

When can the hive stay? When must the hive be moved?

▶ When wild hives establish themselves in and around gardens and yards, they can choose many a varied place. In many cases they choose locations that will not inconvenience a home owner. The hive can have been in residence for many years before a homeowner even becomes aware of it. Examples of this include hives established in brick walls of homes including brick chimneys, elkhorns, concrete block, rock and timber retaining walls, upside down garden pots, bird bath bases and under concrete paths and house foundations. We explain to people just how lucky they are to have these wonderful creatures choose their yard while other people have to pay for the privilege!

▶ In other situations, they choose locations where they cannot stay. For example, water meter boxes, TELSTRA pits, irrigation pits, homes that will not last, (a good example of this is where on two occasions hives had colonised garden potting mix bags with the plastic bags starting to rot away) and homes that have been accidently damaged beyond repair (Image 70).

Image 70 Clear evidence of an active native beehive in a water meter box.

Hive rescue by relocation

▶ When hives have chosen a home where they cannot stay, and that is sound and moveable, our preference is to relocate them in that home with as little disturbance as possible. Relocation could be to a new spot in the same yard or to a completely new location. Examples include upside down garden pots and bird bath bases. This carries little or no risk for the bees and their home can become an impressive garden feature more valuable than a boxed hive (Image 71 & 72).

Images 71 and 72 left to right *A hive found in a 150mm PVC pipe that provided access to an underground water main valve. The pipe and hive were removed to allow plumbers to urgently access the valve. The pipe with its hive were then reburied in the garden to insulate the hive. A fairy house was added over the entry for the grandchildren to enjoy without causing any harm to the bees.*

When does a rescue hive need to be transferred into a box?

▶ If a hive must be moved and cannot be relocated in its current container, it will need to be transferred into a box

▶ Hive rescue can carry significant risks and many first-time rescuers lose their first rescue. Using a well-prepared rescue box, the most important aim is to transfer the brood mass in as good a condition as possible and to get as many bees in the new box as possible. Pollen and honey pots and resin structure attached to the brood mass structure are okay to include but honey sacs should only be included if there has been minimal damage to them. We think leaking honey that was placed in the rescue box with the good intention of providing food for the bees is the main reason for rescue failures. Leaking honey will drown bees that fall into it, ruin any brood that it comes in contact with it and attract pests in the area.

Water meter box rescues

▸ The most common rescue is from a water meter box but you can do a dozen of these rescues without two being the same. The obvious differences will be the species. TCs and THs both colonise water meter boxes with THs by far the most common in South East Queensland. There are big variations in size and volume of the hives and the meter boxes themselves come in different sizes and are made of different materials. There are also big variations in the location of the hive. Some are down at the bottom of the meter box and some sitting high in the box. Some will be constructed around the meter and some under the meter. Some have a lot of vegetation material interwoven with the hive. Some are years old with the propolis set hard and are almost impossible to cut through. Some are so new that the brood is the size of a 50 cent piece.

▸ We set up beside the meter box with a box that has the perspex viewing panel opened to allow easy transfer of the hive; a torch; a sharp filleting knife; a large screwdriver to open the meter box and a damp cloth to wash up afterwards. In spring and autumn, we do rescues about 15 minutes before dawn when all the bees are in the hive, not wanting to fly and close to the brood to keep the brood warm. Doing it this way, nearly all the bees can be transferred into the box. In summer just before dawn the bees will still be in the hive but away from the brood on the walls of the meter box making it harder to transfer good bee numbers. A bee-vac (a vacuum cleaner with a bottle over the end that is screened at the join with the bottle to catch the bees in the bottle) is then useful to catch as many bees as possible (Image 73).

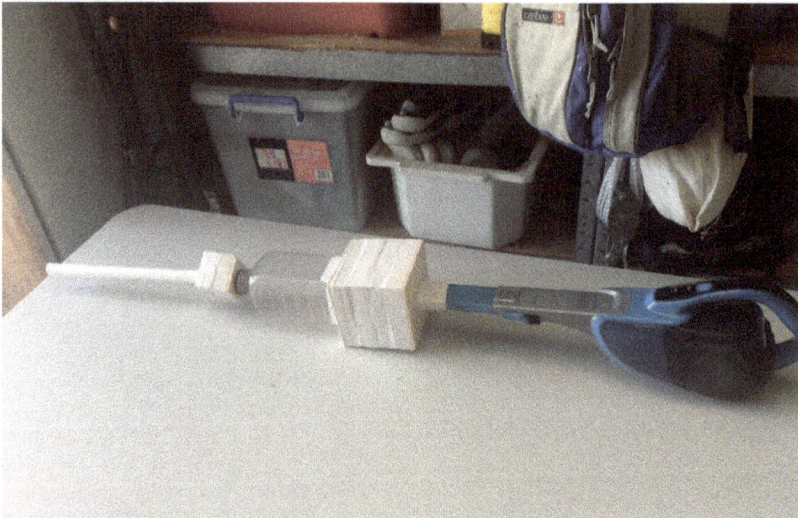

Image 73 *Example of a "Bee-Vac".*

- ▶ We use a sharp filleting knife to cut away the edges of the hive from the walls of the meter box. Separating the hive from the walls of the meter box as quickly as possible stops a lot of bees from crawling out of the hive. A hand is then placed under the hive to lift it out. Any human hand contact with the top of the hive will see large numbers of bees escaping onto that hand.

- ▶ Usually the small hives are able to be removed fully intact, but we find they only have a fifty-fifty chance of surviving (Image 74). If you get the queen with the transfer, the hive will usually survive, but if you miss her or if she is damaged, this young hive is yet to produce queen cells to create a new queen. This is probably because the new hive already has a new queen and needs to produce more workers as a priority. Being small it is probably supported by the mother hive which could provide a new queen if required. It will become evident after a few weeks whether or not the queen has been safely transferred. If not, it is usually necessary to hook this weak hive up to an eduction mother hive.

Image 74 Example of a small TH water meter hive.

▶ Medium size hives are usually the most likely to succeed with the propolis edges still soft enough to be cut with a sharp knife. The hive is large enough to have virgin queens and large queen cells should the queen not be transferred successfully. These hives may have large walls of honey stocks which, if they can be cut away without much damage, can be placed in the rescue box. But if they cannot be removed without damage, it is much better to leave this material out (Image 75).

Image 75 Example of medium size TH water meter hive

- Large old hives are the most difficult to transfer but are also not common. The internal structures the hive has built to support itself can be set so hard that a sharp knife may not be able to cut through it. The brood, particularly a TC brood, may be completely surrounded by honey stores which increases the risk of a significant honey spill. Removing the entire meter box and taking it home for eduction purposes is a good option. The meter box needs to be reburied to provide necessary insulation.

- When you get the hive home, we leave them penned for 48 to 72 hours to give the bees a chance to sort themselves out before being exposed to the big bad world. If there is leaking honey this needs to be drained. We place these boxes under cover where pests are unlikely to venture, and tilt the box back so that the honey can drain out of the hole in the floor at the rear of the box. The drain hole has a second layer of tape over it so that honey does not leak out in the car. It needs to be peeled off and the honey drained into a container for about 24 hours. The honey is a strong pest attractant. A good trick is to place borax in a cup under any dripping honey to neutralise the smell and to continue to wipe away any honey on the outside of the box.

- Winter rescues are particularly risky as recued hives are unlikely to have the capacity to generate and trap much needed heat around the brood. We place rescued hives in locations that will enjoy full winter sun on surfaces that will trap heat like a timber top table. In addition, we insulate the walls of the hive box with foam boards like the products provided by KNAUF Insulation (Image 76 & 77).

Images 76 and 77 left to right *Insulated winter rescue hive on a sunny table.*

Log hive rescues

▸ Firewood cutters, fencing contractors, tree loppers and land clearers occasionally cut down a tree with a bee hive in it and often don't realise this until their chain saw hits the nest and either there is a strong smell of burning honey or a swarm of bees alerts the chain saw operator to their presence. To rescue this hive, you need to act reasonably quickly.

▸ It is best to ask the chain saw operator to cut the log through 50 to 60 cm each side of the entrance hole - using straight cuts to make later resealing easier. If the hive has been cut through, putting logs back into their original positions relative to each other will usually see the hive survive. If the log is fractured it is best to tape around the log to hold it together and to cover and secure the top and bottom of the log using spare clothing and gaffer tape.

▸ If the log is in reasonable condition, knowing that the bees do not need a perfect log to survive, we do not support removal of the hive from the log. A reasonable condition log can usually be turned into an attractive long term garden feature by adding a hardwood floor and roof to the log (Image 78, 79, 80 & 81). Any other saw cuts that may have reached the hive chamber can be filled with No More Gaps. Do not use liquid nails as the fumes are toxic to bees.

▶ If the log is in poor condition (Image 82), or if the hive is located in the root ball, it will usually be necessary to extract the hive and place it in a box.

Images 78, 79, 80 and 81 left to right An Austroplebeia Australis log hive cut into three firewood pieces subsequently re-joined to maintain a thriving hive.
An Austroplebeia Australis log hive cut into five firewood pieces subsequently re-joined to maintain a thriving hive.

Image 82 Tetragonula carbonaria (TC) brood exposed as a decaying log is split open to allow transfer of the hive into a box *by Geoffrey Dutton.*

Hive box replacement

▶ There are several options for managing hives housed in tired old boxes that are rotting away; boxes too poorly made to protect from pests and extreme weather; and boxes built to adhoc dimensions and design which limit, if not rule out, several beekeeping practices (Image 83).

▶ If the hive owner is not seeking to have future splitting and honey harvesting options, the easiest approach is to build a box large enough to encase the old box. Once encased, the hive should be long lived and will usually make an ideal eduction mother hive.

▶ Another option is to transfer the hive into well made, standard size boxes to ensure longevity of the hive and to offer future splitting, eduction and honey extraction options (Image 84).

▶ A transfer into a new box is similar to a difficult rescue from a water meter box. This has the significant advantage is that it can, and should be, planned in advance.

▶ If the box is not so bad that it can be kept for another year or two, put the transfer off until after first propagating at least one hive by either eduction or splitting (where the old box is able to be split). This ensures that you will continue to be a hive owner should the eventual transfer fail.

▶ If the transfer is urgent and cannot be delayed, then proceed in the same manner as you would with a rescue.

*Images 83 and 84 left to right A tired and rotting hive box **by Steve Armstrong**.
The contents of the old box now ready for transfer into a new box **by Steve Armstrong**.*

Managing hive pests

▶ There are three species of pests that native bee keepers need to be aware of, Phorid Fly (Image 85), Syrphid Fly (Image 86) and Small African Hive Beetle.

Image 85 and 86 left to right *A Hive Phorid Fly (Dohrniphora trigonae) caught in liquid trap & Syrphid Fly (Ceriana ornate)* ***by Geoffrey Dutton***

▶ Mature hives in logs and boxes are close to being impregnable to these pests. If the main brood chamber boxes are half full or more, we regard the hive as being mature.

▶ Newly propagated hives from splitting, (the hive with an old top and new bottom) and newly rescued hives are the most vulnerable to attacks from Syrphid Fly (false wasp or wasp mimics), if this fly can gain access to honey leaks; and the Small African Hive Beetle, if they are able to enter the hive. Well prepared boxes and effective management of honey leaks are the most important preventative measures for both of these pests.

▶ Easily the most challenging pest for native beekeepers to deal with is the Phorid Fly - we refer to them as 'horrid' fly. We have observed that small hives and weak hives with spare spaces in the box, particularly near the entry hole, are the most vulnerable.

▶ There are three strategies that can be applied to managing this pest;

• **Strategy 1** - Replace the perspex viewing panel with a temporary timber lid; take the hive into a dark room at night and leave it to settle for an hour. The flies seem to settle on the underside of the timber top. With a torch handy quickly remove the lid and flip it over, now swat the flies as quickly as you can. Repeat this step about an hour later and again the following night until the flies have been beaten.

- **Strategy 2** - Another measure is to lure the flies into a small container of fermenting liquid where they drown. Containers can be placed inside and outside of the hive. The container has holes in it big enough for the flies to enter but small enough to stop the bees from entering. (A suggested mix is – 1 tspn Vinegar, 4 tspn water, ½ tspn honey and just one small drop of dish washing detergent.

- **Strategy 3** - Another trapping and drowning measure is to place the neck of a small bottle (screened so that only flies can enter) into the hive entrance. The flies leave the hive and drown in the liquid (Image 87).

Image 87 *An entry hole Phorid Fly trap.*

▸ The Varroa Destructor mite is an external parasite that only effects European honey bees and is not a threat to native bees. Varroa Destructor has had a massive impact on honey bee populations worldwide wiping out most feral bee hives and many managed bee hives, but thankfully it has not yet made it to Australia. When this mite inevitably makes it to Australian shores, the importance of native bees to commercial and hobby crop pollination should dramatically increase opening up many hobby/business opportunities for native beekeepers.

Image 88 *Native Stingless bees foraging from an Australian native plant Eremophila maculata –* *Pink Passion* **by Geoffrey Dutton.**

Image 89 A beautiful water lily and native stinglees bees at Butterfly Falls north west of Katherine in the Northern Territory *by Alistair Grubb.*

Happy beeing!